Serendipity in Materials Discovery

-

Serendipity in Materials Discovery

(or: For the Want of a Horseshoe Nail)

-

David G. Morris
Madrid,
Spain

Copyright: February 2015

ISBN-13: 978-1507852613
ISBN-10: 1507852614

We live in a world defined by materials. Man lived through the Stone Age, the Bronze Age, the Iron Age, then the Silicon Age. The availability of critical materials has always defined the ways of our societies. On a more specific level, access to a special material, either through pure luck or by good planning, has often made the difference between win or lose, life or death, or simply comfort or hardship. In the following pages we show examples of such availability, or lack, of critical materials that have determined the outcome of certain cases: these have been the nails that saved, or lost, the kingdom.

Materials are what are known, in "Brussels-speak", as enablers. They do not create the machine or the application, but they make it possible to be invented. Without these materials being available the invention will not have been created. Examples of such inventions or occurrences which depend on the availability of the material are shown here.

In the past, materials have often been discovered simply because they were available – the earth had somehow created and offered them – and all that had to be done was to realize how useful the materials could be. Since then, Man has thought himself so clever that he invents new materials – simply by being so clever. Often, however, his cleverness depends on simple luck (also called serendipity), or simple accident, and it is really a question of being in the right place at the right time. Instead of being the clever person that discovers what he set out to look for, he is really the lumbering creature whose only merit is to notice what falls into his lap by accident.

In this short book I describe a dozen or so cases where availability, accident or luck has led to a new material. Each example is discussed in a few short pages to make for easy reading, but also to teach a little of the basics of the science of the materials.

To Leo, Ben, Alex and Lucas
February 2015

Index

1: Napoleon's Buttons:

In 1812 by Illarion Pyranishnikov

Napoleon Bonaparte was probably the best military commander that France ever had. In the years following the French Revolution of 1789 he made France into the most influential nation in the world. By 1812 he controlled all of continental Western Europe and began his expedition to take control of Eastern Europe including Russia. His attempts came to nothing, and in fact marked the beginning of his downfall. Some of the blame for this disastrous Russian campaign may be put down to the buttons holding closed the coats of his soldiers. These buttons were made of tin. In the intense cold of Russia in winter the tin changed to a brittle form and so the buttons fell away leaving the struggling soldiers freezing to death. Impossible to continue, or even return to France, the expedition was a disaster and almost a half a million soldiers were lost. Napoleon lost his reputation as being invincible, and his empire began to crumble.

The French invasion of Russia began in the summer of 1812, with an army of almost half a million soldiers moving into Western Russia looking for a battle against the Russian army. One justification for this invasion was the liberation of Poland from the threat of Russia, and another real reason was to prevent Russian support and trade with the arch-enemy, Britain. This enormous army marched rapidly through Western Russia hoping for a decisive battle with the Russians, but these slipped quickly away burning the villages and towns and denying the invaders of provisions. The French army had to rely on supplies brought in from home, leaving the soldiers hungry and constantly in search of additional food supplies. Even reaching the Russian capital, Moscow, brought no victory for the government had gone leaving no food but an empty, burning capital. Instead, Napoleon was forced to retreat back towards France. But now winter was coming, his troops had no winter clothing, the horses were tired, and food was constantly lacking. It took weeks for the French army to retreat, and the lack of food and the cold led to the starvation of about four hundred thousand men, with many of the remaining being captured by the Russians. The whole affair was a military and humanitarian disaster.

One of the reasons for the disaster was the crumbling of the buttons used to keep closed the coats of the French soldiers. The soldiers had been equipped with thick coats and, in order to reduce the costs of preparing this enormous army, the buttons to close them had been made from tin. During the Russian winter the temperature can drop to as low as -40°C, when the tin transforms to a brittle state which easily breaks, then leaving the poor soldiers with wide-open coats.

The element tin exists in two crystallographic structures, allotropes, or atomic arrangements, near room temperature. Many elements show such allotropy with the commonest, perhaps, being carbon which exists in several forms including diamond, graphite, graphene, and several more forms. The common white form of tin, known as beta tin, is stable above about 13°C and has the atoms arranged in a tetragonal structure – like a cube but with one length longer than the other two. This allotrope is metallic – the electrons are free to move and the material conducts electricity – and is ductile and easily deformed. At lower temperatures, however, the material changes to grey alpha tin with a complex cubic structure, known as the diamond cubic structure. This allotrope is non-metallic – the electrons are not free to move and the material does not conduct electricity – and is hard and brittle. When bent, this allotrope breaks. The change of crystal structure leads to a volume change as great as 27%, which leads to the self-destructive crumbling to dust as the change of phase state occurs, known as pesting. In fact, the pesting process is autocatalytic, meaning that the stresses produced at the onset of the change of phase help neighbouring regions to start transforming and the process accelerates throughout the entire object. In typically impure materials, for example tin containing small amounts of bismuth, lead, or silver, or in tin alloys containing deliberate additions of copper, lead or other elements, the phase change becomes more difficult, and occurs at lower temperatures or more slowly.

Did this embrittlement really happen to the buttons of Napoleon and his soldiers? There is, in fact, some doubt. The temperatures encountered by the soldiers were certainly low enough to start pesting. Tests in the laboratory, however, show that the pesting process takes place very slowly and may need several years to complete at the typical low temperatures – much longer than the several months duration of the Russian campaign. In addition, the material used at this time to produce the buttons would have contained large amounts of impurities which would have made the process even slower. Any doubts seem to be confirmed by a recent examination of graves of some of the poor, dead French soldiers, where legible markings on buttons were used to help their identification. The buttons had been able to retain good integrity for over a century in the cold earth.

So, did crumbling buttons help cause the Russian fiasco for Napoleon? The evidence is not complete, but seems to contradict the legend of Napoleon and his failing buttons. Nevertheless, the legend remains and is still believed by a great many people.

2: Nelson's Bottoms:

The Battle of Trafalgar by Clarkson Stanfield

Admiral Lord Nelson was the commander of the British "Royal Navy" fleet that engaged the combined French and Spanish fleets at the battle of Trafalgar, off the South-West coast of Spain, in 1805. This was an important part of the Napoleonic Wars fought from 1803 to 1815 between Britain and France. Instead of lining up his ships in parallel to the French-Spanish ships, Nelson turned in at the last moment, cutting the enemy line and causing such confusion that he was able to destroy many of their ships and earn a great victory. The critical moment in this plan was the period when the British ships approached head-on since they would be unable to fire their cannons, while the enemy would be able to fire at will. It was critically important to be able to sail in quickly through this dangerous stage: once the enemy line was broken, the British would turn the advantage and be able to fire on the bows and sterns of the enemy ships with no fear of reprisal. The use of copper coatings at the bottoms of the British ships, a modification made over the previous few years, meant that the ships were clean with no worms or weeds, and were much faster and able to sail in more quickly than without coatings.

Throughout the Napoleonic wars, the French had gradually come to dominate the ground in central Europe, but the British had maintained naval supremacy. This supremacy was to be put to the test as the combined French and Spanish fleet approached the British fleet off Cape Trafalgar in autumn 1805. The French-Spanish fleet was much bigger, with thirty-three ships carrying thirty thousand men and two thousand, five hundred cannons opposing the British fleet with only twenty-seven ships, seventeen thousand men and just two thousand cannons. Nelson knew that his sailors were better in their seamanship than the enemy and that his gunners were well-trained, faster and more accurate, but that the battle would be difficult and dangerous with many of his ships being fired upon from more than one ship at a time.

Military convention at that time suggested that the enemy fleets should form two parallel lines, separated just by the range of the cannons, about one hundred metres or so, and then sling cannonballs at each other. These balls weighed up to twenty kilos and, projected at high velocity, could cause severe damage to hulls and break sails and masts. Nelson decided that instead of lining up in parallel he would cut through the enemy lines, causing confusion and allowing the British ships to pick off the enemy one by one. Enemy ships ahead of the intersection would be forced to slowly turn around, taking some considerable time, and be unable to participate in the battle until later, thus making the British advantage even greater.

The critically dangerous stage of this plan was when the British ships approached the enemy line, head-on against broad-side, when the enemy would be free to fire all their cannons but the British unable to respond. It was thus critical to move in as fast as possible over the final hundreds of metres. For heavily-armed ships with a top speed of some five knots (about 10 km per hour) there would be a danger time of a minute or so, but the light winds at Trafalgar meant that the first approach would be longer and could be suicidal.

A major factor improving the speed of the British ships was the copper sheathing used to cover their hulls and prevent the growth of worms, barnacles and weeds. It had been known for centuries that wooden ships were slowly attacked by shipworms, with barnacles (small shellfish) encrusting on to the hull surfaces, and weeds growing down from the hull and steering equipment. These problems meant that gliding through the water was more difficult, and ship speeds fell, but also hull integrity was compromised and the ship could fall apart. An initial solution was to attach a "consumable" wooden overcoat that would be regularly replaced or a lead overcoat sheathing. Lead sheathing led to problems of severe corrosion of the iron bolts used to hold together the hull, and so was not a good solution.

Copper had been suggested as an alternative sheathing material, but its high cost and limited availability meant that it could not be considered for use on a large scale before the second part of the eighteenth century. By this time it was seen that long exposure during transatlantic crossing led to serious structural weakening of the hull and better protection was essential. An experiment was tried in the mid-eighteenth century whereby a British frigate was covered with copper. The protection offered by the copper sheathing was magnificent: there was no woodworm and no fouling by weeds. The copper forms a poisonous oxychloride film in the sea water which deters growth of the marine life, and this film slowly dissolves taking with it any attached growths. Despite this biological success, however, an extremely severe problem of corrosion of the iron bolts was found which made it impossible to use the copper sheathing. At this time, two problems prevented the use of copper as sheathing material: the electrolytic reaction between copper and iron which lead to the sacrificial attack of the iron bolts, and the high cost of copper which would be needed in large scale in sheet form.

Towards the end of the eighteenth century two changes took place. Firstly, with the development of scientific understanding of chemical reactions, electrolysis, and the role of different elements and their relative activities or potentials, for example by scientists such as Sir Humphrey Davy in England, it became clear that the intelligent selection of a better pair of metals – instead of copper and iron – could solve the problem. Iron bolts were replaced by ones made of copper alloy, for example of copper and zinc (brass). Secondly, in Anglesey, Wales, a large-scale source of cheap copper had just been established, that meant that a whole fleet of ships could be sheathed at reasonable cost. By the time of the Napoleonic wars, the British fleet was completely protected against biological fouling by the copper sheathing, and the ships were fast and structurally sound: able to glide through the water smoothly, even with mild winds.

At the end of the day, the French-Spanish fleet had lost some twenty-two ships and the British not even one. 'Twas a great victory", and all with the help of a clean copper bottom.

3: Flintstones:

Sparks from striking flint on steel

Stones made of flint have played a critical role in creating sparks which could be used to make fire. Throughout the Stone Ages, primitive man saw the importance of fire for survival and used tools made of hard and tough flint to strike against an iron-containing alloy or mineral, produce sparks and ignite tinder. Similar impacts of flint on metal were used well into the 19th century, as flintlocks, to ignite gunpowder and fire guns. It was the special properties of the flint, both hardness and toughness, that made these uses possible.

From the beginnings of the Stone Ages, some million or so years ago, man saw the usefulness of fire for attack and defence, for keeping warm, for cooking and for lighting. How to obtain this very useful fire, or how to maintain it, were major problems. Initially, natural fire from lightning strikes could be collected and fed with dry woods to keep alive. Hot embers or coals could be carried, smouldering within an insulating pouch during the travelling day of the nomadic tribe, but this was difficult – and what happened if the fire went out? Techniques of starting fires by friction, rubbing piece of wood against another, were developed, but these required time, patience and difficult skills. How much easier to strike together a piece of flint with an iron-stone, produce a shower of sparks, and quickly ignite the waiting wood shavings.

Within the eroding chalk or limestone cliffs found both on the sea shores as well as inland there were (and they are still found) many small nodules of hard stones of a material called flint. When struck with hammer-stones the flint breaks into thin flakes with sharp edges that have many uses. Not only can these flakes be shaped into cutting knives and axe blades, they can also be used as striking and scratching tools for hitting other stones. Striking flint against itself or other normal stones does not lead to any excitement: but strike the flint against a piece of iron (not available to the stone age man) or against a piece of the common iron pyrite mineral, and a shower of sparks is produced. What better way to produce a quick and easy fire than to direct the shower of sparks onto a bed of dry tinder – in minutes a blaze is obtained. The mechanism is simple: the sharp cutting edge of the flint scratches the iron (or the pyrite), pulling off small particles of metal which are heated by the cutting process itself and which then quickly oxidize in air, i.e. they burn.

Striking a sharp flint edge onto a steel surface was such a reliable way to produce sparks that it was used until recently for making fires or for shooting the flint-lock gun. The flintlock mechanism used a small piece of flint held in the jaws of a spring-loaded hammer. When released by the trigger it struck a piece of steel, producing a shower of sparks which ignited the gunpowder placed just below. So reliable was the mechanism that soldiers and duelers trusted their lives to these sparks into the middle of the 19th century.

Flint is a sedimentary rock, composed of the mineral quartz (silicon dioxide or silica), which forms as small inclusions or nodules inside large volumes of limestone or chalk. It is formed as silica dissolves from other silica-containing sources, collecting as amorphous (i.e. non-crystalline) silica gel in cracks or holes in the parent limestone which then, over long, geological timescales, dehydrates and transforms to crystalline rock. Flint is sometimes considered to have a cryptocrystalline state, meaning simply that its crystal size (or grain size) is so small that it can hardly (or not) be distinguished by eye or by standard optical microscopes. Hence the first people to study it with small microscopes thought that it was not crystalline. In such a state it may be called chert or chalcedony. Being a crystalline form of quartz these forms of flint are very hard, with a Mohs hardness of approximately 7 (on this somewhat qualitative scale that rates the diamond as hardest with a hardness of 10 and soft talc as softest with a hardness of 1). Depending on what impurities are present and on the grain size, the flint can have a variety of appearances – ranging from white or black or brown chalcedony with a waxy shine to duller-opaque chert, also with the variety of colours. While hardness is mostly defined by the crystalline state of the quartz, the grain size strongly affects the fracture resistance of the rock. (Fracture resistance measures how much force or stress is required to propagate a crack, and so has strange units such as Pa-square-root-meter, with the Pascal (Pa) unit measuring the stress and the meter describing the crack length.) Large grained quartz, as often found in igneous granite (the sort of rock that has come out of a volcano), is really brittle and has a low toughness of about 0.5 MPa.m$^{1/2}$, but reducing the grain size to 5-10 microns as found in chalcedony and chert increases the toughness to a respectable value of near 3 MP.m$^{1/2}$. Why is this important? The combination of hardness and toughness is critical for the application, because it means that the nodule fractures along smooth planes defined by the hammer stresses when the nodule is in the workman's hands and a sharp knife-edge can be created. This sharp knife is capable of cutting the metal or pyrite during use without the tool blunting or breaking. Without the special combination of properties - both high hardness and high toughness - the flint would not be at all useful for generating sparks for fire-making but would instead blunt and crumble along the knife edge.

As Stone Age man became more and more clever, he noted that "cooking" his flint, at a moderate 150-250°C overnight (surely this must have been another of those accidental, serendipitous discoveries) led to the stone becoming more uniform in properties and more easily workable. Flakes produced during tool-making now followed the forces imposed by the hammer more readily, and a smoother, sharper and tougher knife could be cleaved off. This was one of the first examples of the manipulation of materials by man - changing properties of the material to suit the final application by applying some processing treatment.

4: Superalloys:

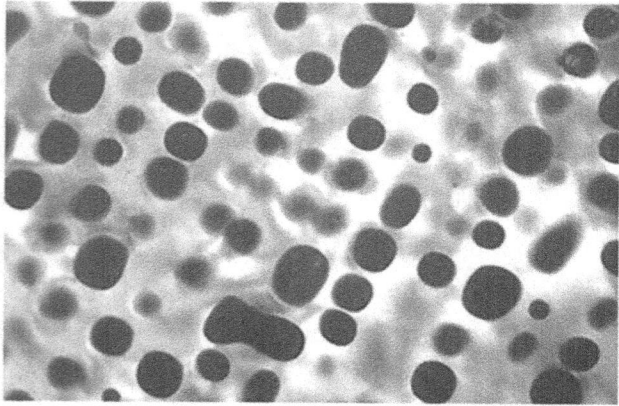

Internal microstructure of superalloy

Superalloys are metallic alloys that are capable of withstanding amazingly high temperatures – very close to the temperatures where they melt – under high loads or stresses and aggressive oxidizing or corrosive conditions. They are used inside the hottest regions of machines such as the gas turbines used to power airplanes and electricity generating stations. These superb materials were invented during and after the second world war as it became clear that conventional piston-type engines driving propellers could no longer be pushed to higher performance and alternative equipment with special materials was needed.

As piston-driven propeller airplanes developed in the first decades of the twentieth century it became clear that the efficiency of the power system was limited, especially as the propeller tip velocity approached the speed of sound. A new power plant concept was required, and this was to be the gas turbine engine, or jet engine. The concept was not new, with one patent dating from the end of the nineteenth century, and involved a compressor providing large masses of air to a combustor where the fuel was burned and hot gases did useful work – turning the compressor, providing thrust as a rapidly expanding high-velocity gas jet, turning a propeller to move a large quantity of nearby air. Experimental gas turbines had been tested from the beginning of the twentieth century, in Norway, England and Germany. There were, however, major problems of material and machine reliability, which meant that gas turbines were not taken seriously for a long time – many of the engines ran for only a few hours before catastrophic failure. Only slowly, first in Germany and then in England, were workable gas turbine engines fitted to aircraft. It was the later stages of the second world war before the British Gloster and the German Messerschmitt jet planes began to fly. Today, jet engines are at the heart of the modern airplane, even most propeller-driven planes, with the same machines found also inside the most efficient gas-burning electricity-generating power stations.

Metals deform at low temperature by the shear of their compactly-packed atomic planes. More precisely, they contain line defects called dislocations on these planes that allow the easy shear of only a small linear part of the shearing plane at one time. For this reason, metals are ductile or malleable – they can be easily deformed to large amounts of shape change – and are relatively soft. The problem for high temperatures machines is that these metals become so soft that they cannot hold shape for any period of time – and they deform by a slow process called creep. The gas turbine engine contains many blades rotating inside a casing or container, with centrifugal forces so great that the blades creep, increasing their length and touching, rubbing, and wearing against the casing wall. High strength and high resistance to creep is required to as high a temperature as possible. The metallurgical mechanisms used to strengthen materials for low temperature usage involve blocking dislocations by any obstacle such that they cannot move easily. Alloying (mixing several metallic species together in the same crystal) leads to dislocation blockage by the foreign atoms, by the agglomeration of such atoms together to form small particles known as precipitates, and by other defects such as the boundaries between the crystals themselves, and by the large numbers of other dislocations produced during deformation. Unfortunately these mechanisms do not work at the high temperatures needed by gas turbines.

The solution to the problem was developed during the second world war, especially in the Wiggin factory of the Inco company in the United Kingdom. The solution – the Nimonic alloys - was to use a suitable metal matrix of nickel where the atoms are highly close-packed, add a large amount (15-20%) of chromium for better oxidation-corrosion resistance, and some 2-5% each of aluminium and titanium to produce large volumes of an atomically-ordered nickel-aluminium or nickel-titanium precipitate. The matrix has been called the gamma phase, with the precipitate having the chemically-ordered gamma-prime phase, and the superalloys are often called gamma-gamma-prime materials. The photograph at the beginning of this section shows the microstructure (seen in a microscope) of such a gamma-gamma-prime material where the cuboid blocks are the gamma-prime precipitates, of nearly a micron in size and occupying more than 50% of the material, and the white inter-block material is the nickel matrix. This microstructure remains very stable (it doesn't change) at the high temperatures inside the gas turbine engine, in part because the atomically flat gamma-gamma-prime surfaces have a very low energy and don't want to be disturbed. Dislocations are forced to remain inside the narrow matrix channels and so have very great difficulty in moving – meaning that the creep strength remains good.

Since the second world war there have been enormous developments to both alloying and processing of such superalloys, as well as the design of the blades themselves. Operation temperatures have increased from 700-800°C to above 1100°C. The temperatures of the gas nearby may be even higher. The base metal can be nickel, or iron-nickel, or cobalt for questions of cost or corrosion resistance; additions of heavier elements such as molybdenum, niobium, rhenium and tungsten strengthen the nickel matrix and the precipitate phase; other additions, including carbon, lead to new precipitate particles that go to grain boundaries and block these. The first superalloys were produced by casting and high-temperature forging; later processing used investment casting to produce the shaped part directly; further high-temperature performance was later obtained by controlling solidification to produce long columnar grains, or even single crystal blades, avoiding the weakening and embrittling effects of grain boundaries as may be found in the usual poly- (multi-) crystalline materials. Finally, the turbine blades are designed to contain channels (holes) that allow cooling gases to circulate, and they are covered with coatings that resist oxidation-corrosion and act as thermal insulation, such that the metal blades can operate at even higher temperatures, in fact reaching close to their melting points.

Developing materials and components with good strength at temperatures very close to their melting is a tremendous metallurgical achievement that has important consequences in terms of energy efficiency, as dictated by Carnot's thermodynamic theorem. Obtaining more thrust with lower fuel consumption is important for aircraft engines. Obtaining more power and electricity with reduced carbon consumption and emissions is equally important. Superalloys are doing a superb job.

5: Age Hardening Aluminium:

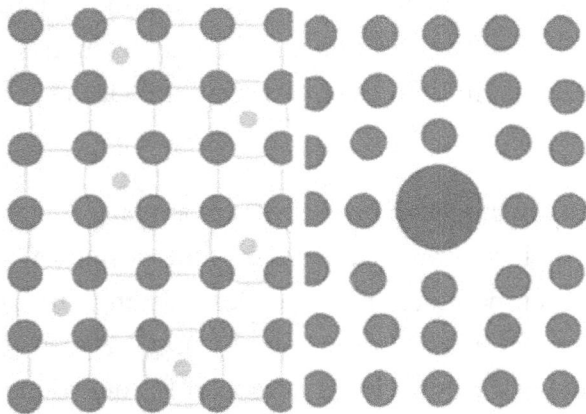

Atoms (small-pink) in solution in Al collect to form a large particle

Aluminium was one of the wonder new materials at the beginning of the twentieth century. It was light and malleable, resisting to many chemicals and a good conductor of both heat and electricity. But it wasn't very strong. It was good for making cooking pans, but not much good for making anything serious.

It was a militaristic age, with the major powers of Europe - England, France, Germany, Russia, and others, spending money on new materials and new equipment – guns, cannons, warships, the new flying machines. The new Aluminium was attractive, especially so light, but it was really not strong enough to be used. In the German military-owned research centre in Neubabelsberg, a young Prussian metallurgist, Alfred Wilm, spent some years examining the possibilities of strengthening aluminium, mostly using annealing (high temperature cooking) and quenching procedures as were common for steels and there effective in producing dramatic hardening. In 1903 this centre was given a commission to attempt the development of aluminium alloys with characteristics similar to brass, used in the manufacture of ammunition, as bullet cases. While much lighter – so a soldier could carry many more bullets – its strength was not adequate and some bullet cases exploded instead of projecting away the bullet itself.

Examining Aluminium-Copper-Manganese alloys, Wilm came close to achieving the required strength, but hardly enough. He had to increase even more the strength. He tried adding small amounts of other elements, evaluating one and another in turn. In 1906 he tried adding a small amount of Magnesium, preparing the alloy by annealing above 500°C and quenching. Still not very strong. But it was the end of the week, so he stopped the experiment on the Saturday morning (at the time working hours were longer than now, and Wilm was a dedicated lad), before stopping for the weekend. There was only time for a few measurements of hardness before stopping. Returning on Monday, he found that the material was much harder. What was wrong? Was it the hardness machine which had somehow become decalibrated? No. In fact it turned out that the material became progressively harder over the coming days. The phenomenon of age hardening had been discovered.

Wilm repeated the tests, confirmed the effect, and developed understanding of the best alloys and best heat treatments to obtain large degrees of hardening. He took out a patent on an alloy with about 4-5% copper together with about 1% magnesium and some Mangenese in the aluminium matrix. In 1908-1909 this alloy became commercially available under the trade name Duralumin. In a twist of fate, the German army showed little interest in the alloy so Wilm bought back the rights to the alloy and sold these to the British Vickers company who used it for the construction of a balloon airship. In yet another twist of fate, the English Mayfly balloon broke into pieces during testing and the strong but light alloy was abandoned, only to be quickly taken up again by the German Count von Zeppelin for his "Zeppelins" - airship balloons. About 100 of these were constructed during the first World War, requiring almost 1000 tonnes of the new material. The same alloy was later adopted by Professor Junkers who produced the first aluminium passenger plane in 1919.

Understanding the age hardening phenomenon was difficult at the beginning of the twentieth century. The only microscopy available was optical microscopy, limited by a resolution of ¼ to ½ micron, and was incapable of detecting any change of microstructure – the inside structure – of the alloy. It was only in 1937 that the French researcher Guinier and the English researcher Preston independently proposed that small groupings or clusters of copper-rich regions formed during ageing. Such internal changes led to internal stresses and strains that were visualized as streaking in diffracted x-ray beams, with these same stresses-strains leading to material hardening. The basic process involves collecting together the randomly distributed (copper) atoms from an initial solid solution in the aluminium (illustrated in the left-hand part of the figure) such that these atoms form a few larger precipitates (illustrated in the right-hand part of the figure). Such precipitates remain very small, only about 1 to 30 nanometres in size when maximum hardening is achieved. In the same way, even though the precipitates are much more widely separated than the initial solution atoms, they are still only some 100 nanometres apart and capable of achieving high levels of hardness.

Today, aluminium alloys hardened using the same principle as that discovered by Wilm are still the main material used for aircraft fuselages (at least until carbon fibre composites may eventually take over completely). Alloys such as those called "2000-series alloys", containing about 4% Cu and 1-2% Mg, and those called "7000-series alloys", which contain about 7-8% Zn, 2-3% Mg and 1-2% Cu, are the most used, and are much improved versions of the original Wilm Al-Cu-Mg alloy. Such age hardening alloys are almost universal in metallurgy, however, with an enormous number of copper-base, iron-base, nickel-base, and other alloy systems, hardened using exactly the same phenomenon.

6: Fluid Bronzes:

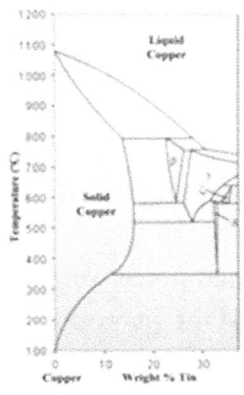 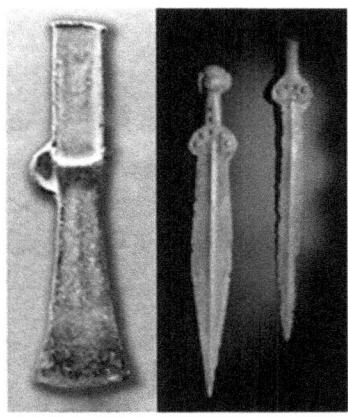

Shapes in Fluid Copper Alloys

The Bronze age stretched from about 3000 BC to about 1000 BC, and during this time smelters and metallurgists learned how to make metals from the mineral ores and to make complex shapes with the desired properties of strength with toughness as well as colour and shine. At the very beginnings of the Bronze age, it was so difficult to manipulate the bronze alloys that archaeologists have defined an intermediate age, called the Chalcolithic age, stretching back about 2000 years before the Bronze age, when essentially stones and copper pieces were used together, side by side – the metal being native copper, that is chunks of pure element occurring naturally. Small pieces of such copper would probably have been heated to soften and hammered to join together into larger pieces, which could be shaped by such hammering. It was not possible to melt the copper to the liquid state since this required high temperatures (the melting point of copper is about 1085°C), and in any case such molten copper is very viscous (like treacle) and very difficult to pour into shaped moulds.

Bronze is a mixture of the element copper with other metals, especially tin, and is much stronger than the metals individually. Depending on the additions, and how much, the bronze alloy can be strong and at the same time be ductile or brittle, so it is important to control these additions.

Choosing the correct additions to the copper base metal allows this control of final properties, but brings with it the important added advantage of melting at a lower temperature and leading to a much more fluid liquid metal – much easier to achieve with the rudimentary furnaces of the time and much easier to fill the complex shapes for jewellery, tools, weapons.

As furnaces became more powerful, so higher temperatures became possible. Much of this development of technology took place in the mountains of the Middle East (Anatolia, Sinai) where mineral ores were available and the winds in the mountainous regions fanned the flames to generate higher temperatures. The smelters of the Chalcolithic-Bronze age began to smelt copper minerals to extract copper metal. They worked with minerals such as malachite (a copper carbonate) and copper sulphides, which they mixed with woods or charcoals to reduce the mineral to the metallic state. Luckily for the early smelters, their copper ores were impure, and contained large amounts of elements such as arsenic or antimony, sometimes even tin.

Luckily for the early smelters, the presence of such impurities brings a number of benefits. Firstly the melting temperature is reduced greatly – to temperatures of the order of 950°C. This is an effect like adding salt to water or ice, which then remains liquid down to lower temperatures. In fact, as indicated in the left hand side of the figure at the top of this chapter, the alloy begins to melt at a temperature as low as about 900°C but only finishes solidifying near 1000°C. At temperatures between these two extremes the metallic mass changes from being a thick liquid to being a soft solid – it can be very easily worked by hammering as it slowly cools. Also, when liquid, the alloy is very fluid – more like water than thick treacle – so it can be easily poured into a container or mould to take up a complicated shape. Finally, on top of all the advantages of easier processing, the alloy is much stronger than copper, and will not be brittle unless there is too much impurity. So the cast necklace, axe or sword will not bend, blunt or break.

It was pure luck that early copper ores were rich in impurities, especially arsenic, and this allowed the transition from the Chalcolithic to the Bronze age. The natural processes that had collected copper into metal-rich ores had also concentrated arsenic, antimony, tin, etc. Without this lucky occurrence, the transition to a metal age with complex-shaped objects with excellent strength-toughness-sharpness would have been greatly delayed. Unfortunately for the smelters, working with arsenic-containing alloys, and in an atmosphere containing arsenic fumes, was not a healthy task. Arsenic poisoning develops slowly over many years, but inevitably damages nerves and internal organs. Initial headache and drowsiness leads to blindness and eventually coma and death. Such work-place hazards led to the replacement of arsenic by tin, which gave similar processing and metallurgical alloying benefits but without the health problems. This replacement of arsenic relied on the availability of suitable tin-containing ores, and in their absence (as in Egypt) the poor smelters carried on suffering. Indeed, it may be the agony suffered by these pre-historical smiths that has been passed down in the legends of the Greek and Roman smith gods (Hephaestus and Vulcan, respectively) who are both represented as being lame – they had suffered from the excess of arsenic at work!

Over the millenia of the Bronze age, metallurgists developed a deep understanding of how to modify the properties of their bronzes by the selection of suitable additions, tin especially, but also antimony, nickel, zinc or other elements. They didn't know the chemistry of their alloys but they knew that ore from this source, combined with ore from that source, and another pinch of that ore, would give them the final material they wanted. Smelters and casters learned how to manufacture intricate objects, especially after the development of split-mould and precision investment-casting technologies. Smiths learned both how to forge shapes and to control properties in these bronzes.

Even today, the casting of copper alloys remains as much an art as a technology, with alloys being classified into several groups according to their castability. Those copper alloys with a small freezing range (with a difference between first melting and final melting of, say, less than 50°C) are difficult to cast since the metal may freeze before filling its mould.

Copper alloys with an intermediate temperature range (50-110°C, the case for many brasses and bronzes) are easy to cast, with the metal being liquid enough to reach the deepest holes before solidifying. On the other hand, those copper alloys with excessive freezing ranges (110-200°C, as for red brasses and some bronzes) suffer from cracking during casting since some metal may remain still liquid when external mechanical or temperature-induced forces come to bear.

7: The Discovery of Graphene:

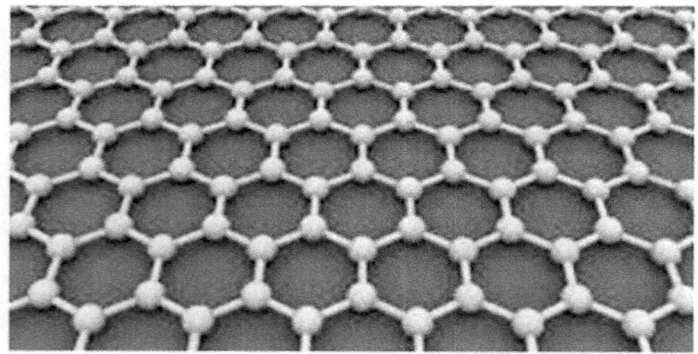

Carbon atoms in Graphene

Graphene is one of the wonder materials of our time. It is strong yet flexible, conducts both heat and electricity amazingly well, and can be doped with foreign atoms to give it special electronic properties and chemical sensitivity. It is claimed to be as important for the future as silicon is to the age of semiconductor electronics. Graphene was predicted to be an outstanding material a long time ago, but could never be made despite many attempts, until a new research student was told to get out of the way and try some pretty useless experiments.

Graphene is one of the many allotropes of carbon – that is, it is a material with the atoms arranged in a specific way, different from that found in the other allotropes (such as diamond, graphite, nanotubes, bucky-balls, and so on). Diamond has the four outer electrons of the element carbon arranged in a strong three-dimension arrangement (technically these are known as sp3 bonds, and this arrangement ties up all the outer electrons of the carbon atom such that diamond is not a conductor of electricity). Graphite has three of the four electrons tied in a flat planar sheet (technically these are known as strong sp2 bonds) with the fourth electron loosely attached. These planar sheets are loosely stacked together, like a pack of cards, with only weak electrostatic van der Waal's bonding between them. The three-dimensional material (pack of cards) conducts electricity well because of the free, fourth electrons of the carbon atoms. These planar sheets of graphite can slide easily over each other, again like cards in a pack, and hence graphite is used as a solid lubricant, and leaves traces on paper as a graphite pencil point slides over the rough material. Graphene is simply an isolated, single layer of graphite, and so its properties are dictated by these strong in-plane sp2 bonds and the free, fourth electron. The bonds of the carbon atoms form a flat hexagonal array, illustrated in the image above.

Graphene had been investigated theoretically for the best part of a century, but no-one had been able to make it. In essence it seems simple – take a flat plate of graphite and thin it by scalpel sectioning or by polishing until only one atomic layer remains. Easily said, but impossible to achieve. Professor Geim at the University of Manchester in England was intrigued by this theoretical material and suggested to a new doctoral student that he continue polishing-thinning to see how thin could be managed. Why not tear off debris from a graphite pencil with Scotch tape and see what it looks like? This idea was taken seriously by a colleague, Kostya Novoselov, who decided to try this experiment in his spare time. Within ten years Professors Geim and Novoselov were awarded the Nobel Prize in physics for their preparation and analysis of graphene.

Graphene remains one of the strangest materials known, and an enormous academic and technological effort is now dedicated to its fabrication, to understanding its properties, and looking for applications. It is incredibly strong, for its thickness, because of those strong sp2 bonds – comparison is made with the strongest steel, with graphene being "hundreds of times stronger". It is claimed to have a strength as high as one Tera-Pascal (a million, million Pascals) , with steels reaching only ten Giga-Pascals (ten thousand million Pascals) at their best, but of course the extreme thinness of graphene pieces means that the loads that can be supported remain small. Stretching the sp2 bonds allows elastic deformation (stretches when pulled and returns to starting shape on unloading) of up to 20%, so the material can be pulled or bent without breaking. Typical metals deform plastically (i.e. don't return to starting shape on unloading), while ceramics break, at much smaller strains. As mentioned before, it conducts electricity well, much better than copper, because of those free electrons, and also can conduct heat in a similar way. Conductivity is in fact amazingly good since the charge carriers are described as having no mass and move at speeds approaching large fractions of the speed of light. It has other special physical properties – such as absorbing a small fraction of transmitted light and hence it can be seen by eye, and anomalous Hall effects (the creation of a transverse electrical voltage when an electric current passes through it at the same time as a perpendicular magnetic field is applied). It also shows great chemical sensitivity to ions or molecules (such as metals, ammonia, hydroxyl radicals) that attach to its surface or insert themselves into the carbon lattice – thus offering interest as chemical sensors.

Intense activity over the last decade has seen the development of many methods to manufacture graphene. When pure (or carefully doped) graphene layers are required, as for electronic applications, methods based on chemical vapour deposition have been developed – these methods are well understood and applied in many industries, for example in silicon chip manufacture or in coating machining tools, and will be readily adopted for future graphene fabrication. For many other applications where demands for purity or crystal perfection are not so great, as for many mechanical applications, techniques for producing large quantities are still under development but are also already in production. These methods are based on industrial comminution techniques, such as grinding graphite with balls or cutting blades – much like a kitchen blender – and probably using some surface activator to help the separation of the graphene layers from the initial large crystals of graphite.

Such techniques are already successful, and graphene is already sold commercially in special inks and paints, which cover a complete surface or a defined pathway with a highly conductive layer. Around the corner, probably, is the use of enormous quantities of graphene flakes (possibly as single atomic layers but probably as a distribution of multilayered flakes) as strong reinforcements inside metal or polymer matrices – something like the carbon fibre/epoxy composites used today to make aircraft fuselages. Most exciting will be the applications in electronics – transparent, wear-resistant, bendable, conductive smart-phone screens, chemical or biological sensors integrated into a semiconducting device, heat-extracting substrates of micro-miniaturized electronic devices.

The long term impact of this revolutionary material in our society will surely be enormous. And it all follows from that one simple experiment – go off and do something out of my way, with a pencil and sticking tape.

8: Invariable Dimensions:

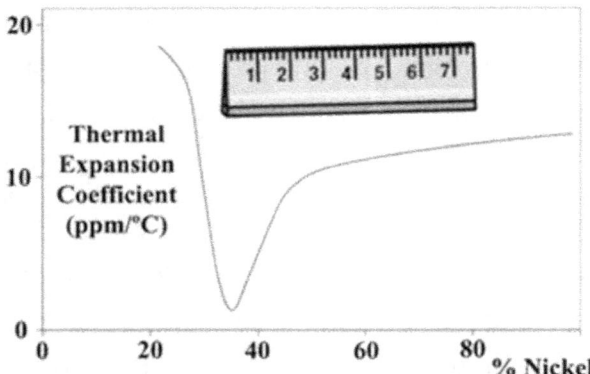

Low thermal expansion alloy

There are many occasions when we need an object to have a precise size. Examples are the standard meter ruler or the pendulum of a clock. If I want to know the exact length of an object (exactly how big is that gold bar I want to buy? How far did that athlete jump?) or if I want to make a clock that keeps perfect time with its pendulum of precise size, then I need my reference measuring ruler to be exactly a given length – ten centimeters, one meter or whatever. But also I need that ruler to be correct on a hot summer's day and an icy cold day. So I need to make my ruler from some material that doesn't change with the temperature – it must have no thermal expansion. This was a big problem for the International Bureau of Weights and Measures – the organisation set up in 1875 in France to maintain standards – as well as for mechanical watch makers. Neither rulers nor watches were standard when taken to tropical countries or to polar regions.

The thermal expansion of metals and alloys was examined in the late nineteenth century by the Swiss scientist Charles Edouard Guillaume. Born in Neuchâtel in Switzerland, the son of a watchmaking family, he was, from a very young age, acutely aware of the problems of thermal expansion – an object increasing in length at higher temperatures. He was employed by the International Bureau in Paris and told to study the fields of mechanics, ballistics and thermometry, before beginning a study of the effect of temperature changes on object dimensions. Alloys based on Iron and Nickel were the usual materials used for making precision parts like rulers, pendulums or other instruments since these alloys were strong, inexpensive and resistant to rusting. So he began to study the thermal expansion coefficient (how much dimensional change per unit of temperature change) on iron-nickel alloys with varying amounts of nickel. The elements iron and nickel have very similar expansion coefficients, so mixing the two together should produce no great differences. Imagine his surprise when he discovered that alloys containing near 36% nickel showed virtually no change of dimension as the temperature changed. Ordinary steels have thermal expansion coefficients of about 10-20 parts per million per unit temperature change (x 10^{-6}/°C), while some of these invariable iron-nickel alloys had values approaching a value of 1 x 10^{-6}/°C. (Note that a thermal expansion of only 10-20 parts per million for one degree of temperature difference doesn't sound like much, but it would lead to a 1-2 millimeter change in a meter ruler when used in the tropics or the poles.) Guillaume's surprise led to the patenting in 1896 of the new commercial alloy Invar (invariable dimension) alloy, and later, in 1930, to the Nobel prize in Physics.

 These invariable alloys are essentially solid solution materials, meaning that the iron and nickel atoms are randomly mixed in the solid state with the crystal structure being the same face-centred cubic structure as nickel. To obtain this atomic mixture and crystal structure it is necessary to heat the alloy to about 800°C and rapidly cool (for example by throwing the object into water), and follow this by annealing (soaking) at a mild oven temperature of 300°C to relax out any thermal distortion stresses. The material obtained has a near-zero expansion coefficient for the small temperature range between about 0°C and 100°C – the usual range where instrumentation is used. As seen in the graph at the start of this section, the very low expansion phenomenon over this temperature range occurs only for alloys within a narrow composition range, near 36% nickel.

The reasons for the strange expansion behaviour are not easy to understand (and in fact research continues to this day to better understand what is going on) and are not directly related to the atomic structure, which remains unchanged over the small temperature and composition ranges. Instead the anomalous expansion behaviour is related to a change in the magnetic configuration, with this change bringing about a lattice contraction that almost exactly compensates for the normal thermal expansion. It should be remembered that iron and nickel are different magnetically, with iron showing a strong level of magnetism (about 2.2 Bohr magnetons) to high temperature (800°C) and nickel a weaker magnetism (about 0.6 Bohr magnetons) to a lower temperature (600°C). It appears that the Invar alloys possess an electronic structure (hence magnetic behaviour) that is sensitive to the interatomic distance. The iron atoms can exists in two magnetic electron-spin states (think of magnetism as due to the charged electrons spinning like a top) – a highly magnetic state with a larger crystal size, and a weaker magnetic state with slightly smaller crystal lattice parameter. At the correct temperature for the given alloy, a suitable change of temperature or composition induces a slight lattice expansion which leads to a change of magnetic state, and this in turn compensates the crystal dimensions by inducing a negative lattice expansion. The effect is somewhat similar to a positive thermal expansion being compensated by a negative magnetic-striction as the internal magnetic intensity decreases.

Over the century since their discovery by Guillaume, controlled expansion alloys have been greatly studied and other variants developed. The Invar alloys are well known for their low, controlled thermal expansion. The Elinvar alloys, containing a similar amount of nickel with some chromium in the iron base, have instead a temperature-independent modulus of elasticity, meaning a constant spring vibration, as well as being nicely corrosion resistant and non-magnetic. These are all wonderful properties for precision oscillators, e.g. in watches. More recently controllable thermal expansion properties have been developed in a range of titanium alloys, with behaviour seeming to depend on other electronic changes instead of changes of the magnetic characteristics.

Today's world seems to be dominated by electronics with little care for precision metallic materials. It might be thought that such "old-fashioned" metallic alloys as discovered somewhat serendipitously by Guillaume over a century ago might find little interest today. Nothing could be further from the truth. Measuring devices, such as the pendulum of a clock or the balance wheel and escapement mechanism of the modern Swiss watch as well as the precision micrometer, still need metallic parts to feed the raw data of time or position into the digital display which is the interface with the human user. Controlled, or zero, expansion alloys remain critical for such applications. Joining metal to glass, to ceramic or to silicon – as in a light bulb, a laser, at an electronic substrate or connector – is also an essential part in the manufacture of these devices where differences of expansion between materials must be limited to avoid fracture or other failure. Precision metallic alloys, such as those discovered by good luck by Guillaume, are here to stay.

9: Combinatorial Materials:

Randomly Permutating Combinations

In most of the examples discussed, materials have been discovered by luck, or there has been some element of randomness in properties due to some accident of material availability or of processing method. Most of these examples also date back several years or even several millenia, as such before our modern scientific age. Today, there are attempts to provoke such random discovery by developing test methods such that very many random combinations of material chemistry or structure are tested at the same time. This is the modern field of combinatory materials development.

The field was first used by the chemists, often working in pharmaceutical industries, who used it to examine the multiple variants of chemical compounds that could be imagined to have useful properties for health treatments. The principle is simple: imagine a molecule made of three parts 1-2-3, where each part may be any of N_1, N_2, or N_3 sub-parts, and so there are $N_1 x N_2 x N_3$ possible permutations of molecule possible; then we will manufacture all $N_1 x N_2 x N_3$ combinations and check to see whether they have the useful properties we are looking for. It sounds easy, but just imagine that we have 50 varieties of part 1, another 50 of part 2, and several hundred of part 3 – we have a million combinations!!! Blindly (but with the help of robots and automatic property sensing) we will make up all these combinations in a three-dimensional array – and hope for luck. It's a bit like making one hundred pizzas with ten levels of pepperonis (none, one ... to too many), permutated with ten levels of olives (none, one .... to too many) and finding the one that we really like. A lot of work, but we are sure to find the really best one. All you have to do is eat all those pizzas!!

Following the success of such techniques in the chemical/pharmaceutical industries, attempts have been made to develop new materials using the same methods. In some ways metallic materials are easier to handle in that these are often simple combinations of a relatively small number of elements and each element itself is a simpler object (imagine a spherical atom of given size or charge) than the pharmaceutical sub-molecule which may have a complex shape and uneven distribution of electrical charge or dangling chemical bonds.

One of the easiest ways has been to deposit elements one by one, with a gradient of composition of each element in a different direction. Such deposition techniques are simple to use and highly controllable – for example electro-deposition or evaporation-deposition. We can, for example, deposit element A with a gradient from none to a lot in one direction (say left to right), then deposit element B on top with a gradient (again from none to a lot) in a different direction (say from front to back), and then deposit element C on top with a few levels (little, some, much) in closely-spaced but distinct small regions repeated over the whole two-dimensional array. All we have to do is mix these atoms together, for example by heating in a furnace, and we can test all permutations of AxBxC compositions. The analytical method is even more powerful if we remember that each component (A,B,C) can be a mixture of elements instead of being only one simple element.

This process has indeed been carried out successfully for the development of new varieties of several materials, including some with luminescent properties (producing light for some chemical-electronic reason, not because it is heated to very high temperatures) or for catalysts (materials that activate some chemical reaction to a higher speed without actually taking part, or being consumed, in the reaction).

Other studies have tried to use the same deposition techniques to identify better combinations of battery materials - which must work together as anode-electrolyte-cathode components. There is enormous pressure to improve batteries for our mobile phones, laptop computers, electric cars, and so on, and great progress made over the past decade with very much more yet to come. In this case, however, a serious problem has been found of difficulties of scaling materials and properties from those produced by the deposition experiment to the materials obtained when manufactured on a larger scale. Different methods may instead be used for preparing samples for the combinatorial experiments with their suitable chemical gradients. As one example, combinatorial studies have been carried out depositing metallic powders, instead of individual atoms, in a way that can be readily up-scaled towards industrial fabrication, but this example does illustrate one of the limitations of the whole philosophy when applied generally to metals and alloys.

One of the reasons that metals and alloys are so difficult to invent (or discover) is that their properties depend as much on their chemistry as on the number and varieties of defects – imperfections of all types – inside the crystals that make up these materials. Indeed the examples that have been described above – luminescence, catalysis, even many battery properties – are ones where behaviour depends strongly on the average chemistry and much less on imperfections. For most of the other materials described here – Napoleon's buttons, superalloys, age hardening aluminium, and many others – the important properties depend very sensitively on imperfections present, with the chemistry of the material only providing the base from which such imperfections can appear. For example, Wilm's hardening aluminium needed the rearrangement of atoms in a specific way – needing time and temperature – to occur; if he had somehow lived on Mars or on Venus he might never have found age hardening.

Materials, in general, may be too sensitive to too many parameters in addition to chemistry for the combinatorial approach to be of general use. Exactly how the material is manufactured is important (for generating imperfections at the same time) as well as how and where it will be used (determines which imperfections are relevant and how they may change). In this case, we are still limited by "lacking horseshoe nails" and still waiting to be surprised by new, accidental discoveries.

10: Stainless Steel:

Stainless Cutlery and Brearley

Man has been using metals and alloys for thousands of years. Iron alloys began to be used about three thousand years ago as techniques were developed to extract the metal from iron oxide ores. One of the most important properties of the iron alloys was that they could be made very hard by a heating and rapid cooling (quenching) procedure. At high temperature carbon impurity atoms dissolve in the spaces (interstices) between the iron atoms, cannot move away as the material is quenched, and so leave an atomically distorted, very hard material. Such iron alloys and steels were used for swords, then railroads, sky-scraper beams, and so on. The only big problem with this material was that it rusted. Left for some time in naturally humid air the material corrodes – and that shiny sword or railroad track becomes a sad brown, flaking object. Until one day in 1913 an English metallurgist discovered, accidentally, an iron alloy that simply didn't corrode at all – stainless steel had been discovered.

By the beginning of the 20th century normal steels were used for making gun barrels. Don't forget this was a military age, when tensions between countries in Europe were leading many to improve weapons and defenses. One of the problems of gun barrels was that they rusted – or more correctly they eroded because of the hot gases produced when gunpowder exploded to propel the bullet out. This was a problem because it meant that the gun no longer shot so fast or accurately after firing a few times. One of the people given the task of finding better steels was the English metallurgist Harry Brearley, research leader at the Brown Firth laboratory in Sheffield. In 1912 he was told that the erosion problem was one of hot corrosion by the burning gases and that he should look for a steel that remained strong at high temperatures. He examined many iron alloys containing small amounts of chromium and carbon since he knew that these elements would form Cr-C compounds (carbides) that would give the steel strength at high temperature. He spent a long time looking at these alloys, but none seemed to be sufficiently good for the gun barrels. The story is told that he simply threw into a pile all these uninteresting alloy pieces. A long time later, in 1913, he noticed that one of the alloy samples remained shining in a mess of otherwise rusting steel pieces – unwittingly he had discovered a non-rusting steel, or "rustless steel" as he called it, later named "stainless steel". (A different version of the same story tells that he carried out routine metallurgical studies of the microscopic-structure (microstructure) of each alloy, a procedure that involves polishing and then attacking slightly with a dilute acid, and was struck by the difficulty in attacking some of these new alloys.) His alloys containing above about 11% chromium with only a small amount (0.1-0.2%) of carbon were remarkably corrosion resistant.

Ordinary steels rust as the oxygen in air (or in water) reacts with (oxidises) the iron to a red-black oxide scale which is cracked or porous and then more oxidation can take place. The chromium in the new stainless steels is more reactive than iron such that a thin layer of a chromium oxide is formed on the steel instead of the iron oxide. This chromium oxide layer is very thin, only some microns thick, but completely covers the steel to prevent more reaction. If ever the layer is broken, by scratching the surface or bending the steel object for example, then a new micron layer forms quickly, healing the object and preventing further attack. In this way the material remains stainless for ever, irrespective of usage.

Brearley was born in 1871 in the town of Sheffield, England. Sheffield was well known as a place for making steel, and indeed Brearley's father was a steelworker. Brearley himself left school when twelve to labour with his father. A determined lad, he studied at home and in night schools such that in 1908 he was made head researcher of the Brown Firth laboratories. As often in scientific research, much of the groundwork behind the new discovery had accumulated over the previous years. As far back as the 1820's the English scientists Stoddard and Farraday and French scientist Berthier had noted that adding chromium to iron led to resistance against attack by acids. Towards the end of the nineteenth century the Englishmen Woods and Clark patented an iron alloy containing lots of chromium with tungsten as a very hard and corrosion resisting material, and the Frenchman Brustein emphasized the importance of low carbon content (to avoid the formation of chromium carbides and a nearby region poor in chromium) in high-chromium alloys for good corrosion resistance. Understanding was thus accumulating over a century or so, but it was only the consistent study by Brearley, who altered both chromium and carbon contents, that allowed a clear definition of which alloys would be corrosion resistant.

Despite the invention of his stainless steel, Brearley was unsuccessful in solving the gun barrel erosion problem since his alloys were not hard enough. Knowing that cutlery (made in Sheffield) suffered from corrosion, so that silver knives were used by the wealthy, he proposed that his new steel be used instead. Unfortunately his alloy, with its high chromium content and low carbon content, was corrosion resistant but not hard enough to retain a sharp, cutting edge. He became known as the man that had invented the knives that wouldn't cut!!

Since that time many stainless steels based on iron-chromium-carbon have been developed and find an enormous range of uses. There are three main varieties of stainless steels: ferritic, martensitic and austenitic. These names refer to the body-centred cubic atomic structure (of ferrous/iron alloys), to the martensitic structure (a highly distorted body-centred cubic structure) and a face-centred cubic (austenitic) structure. Ferritic stainless steels have about 11-15% of chromium with little carbon (about 0.1%) and are only moderately strong, but relatively cheap and easy to produce. Martensitic stainless steels have a similar chromium content with more carbon and become very hard when rapidly cooled from red-heat: they are good as cutting knives. Both these varieties are magnetic (ferromagnetic - as ferrous steels) with reasonable corrosion resistance. The austenitic stainless steel has more chromium (15-26%), with about 10-30% nickel that ensures a face-centred cubic atomic cell. This material is highly corrosion resistant, but is not very hard, so it makes good stainless cutlery, but not sharp cutting knives. It is not ferromagnetic, has good toughness (doesn't break easily) even at very low (arctic or liquid-gases) temperatures, and also has good strength to high temperatures. It is therefore used in very many cold, moderate temperature, and hot places, and where non-magnetism may be important - such as pipes for liquid gases, steam tubing for heat transfer from boilers, household cutlery and medical equipment.

11: Let the Light Shine: Tungsten Filaments:

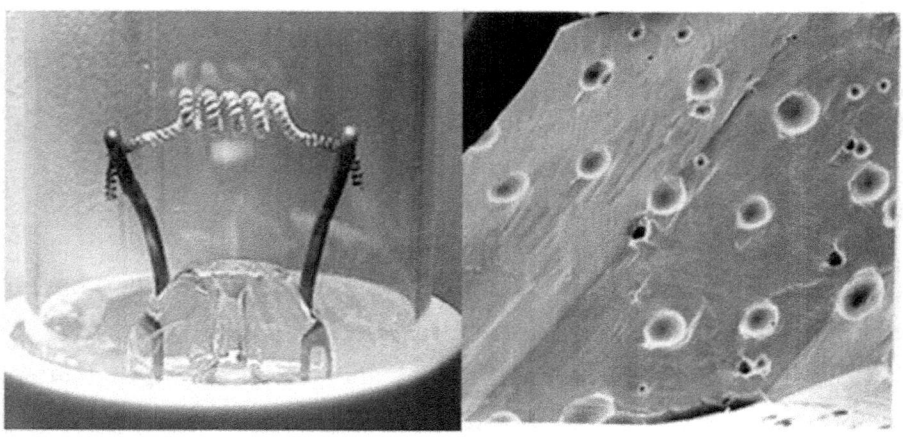

Tungsten Filaments and Bubbles

Near the end of the nineteenth century electricity was seen to be something that would have a wide range of applications. Thomas Edison announced that he had developed an electric light bulb that was sufficiently good to provide more intense lighting than candles and gas lamps. People like Volta and Davy had demonstrated glowing filaments as a light source near the beginning of this century, but these filaments broke after a very short time. By the 1880s Edison had managed life-times as long as 13 hours with carbon filaments protected under vacuum. After many experiments with heavy metal filaments, Coolidge finally developed a new tungsten material that was ductile enough to be drawn into fine wire with excellent structural stability when heated to white-heat inside the electric lamp.

The problems faced by Edison and others in the late 1880s with their heated filaments were many. The filaments had to be long and thin to provide a high electrical resistance, so the electricity would heat them when using reasonable voltages and currents. Oxidation would take place at the high temperatures, so the filament had to be protected within a vacuum or inert gas chamber. Carbon had a high vapour pressure at the filament temperature, and so slowly evaporated away, changing the electrical resistance and darkening the inside walls of the vacuum chamber. Manufacturing a long and thin filament was not easy, since all the materials considered were brittle and ensuring reliable filaments was very difficult.

The carbon filaments were prepared from a variety of starting materials, for example from cotton or thread, from wooden splints, and finally from bamboo filaments: these were heated under vacuum (without burning) to transform the material into a brittle carbon filament. In addition to being brittle, these carbon filaments suffered greatly from the evaporation problem, so filament temperature had to be kept relatively low and their energy efficiency was terrible - only about 3.5 lumens per watt (amount of light per electrical power). Later attempts were made to use more ductile heavy metals with very high melting points, such as platinum, osmium, tantalum and tungsten. These could all be heated to higher temperatures, without such problems of evaporation, and so the energy efficiency was higher. However, they were all expensive and difficult to process. Edison proposed to use expensive platinum, but needed a complex power regulating system to avoid overheating and failure - so the lamp would switch itself off from time to time. Austrian researchers began to use osmium with some success - a light efficiency of 5.5 lumens per watt - but the filaments were so difficult to make that they were not reliable and they were expensive. Tantalum filaments were developed in Germany with some success; for example, the steamship Titanic was fitted out with this modern tantalum filament lighting. Despite this progress, however, there were many problems – the osmium suffered from evaporation loss and the tantalum wire filaments were brittle, especially when used with the now-fashionable alternating currents. As of 1903 filaments were developed from the metal tungsten, with much higher light efficiencies of about 8 lumens per watt, but this material was so brittle that manufacturing was difficult, and filaments were prone to breakage. Nevertheless, by about the year 1910, such brittle filaments had come to dominate the lamp market.

A breakthrough came in the years 1908-1909, when Coolidge at the General Electric research laboratories (set up by Edison and his Electric Light Company) discovered a way to make ductile tungsten. Coolidge realized that tungsten was not intrinsically brittle, but was embrittled by impurities such as carbon, hydrogen and oxygen. Tungsten was prepared by the chemical reduction of oxide (heating it under hydrogen) and the source of oxide played an important role in determining the final properties - sometimes the final material was ductile and other times brittle. Trial and error showed him how to ensure ductility. On top of better ductility, he was amazed to find that his tungsten filaments were much stronger at high temperatures and did not fail by grains sliding one over another leading to a sort of sheared-filament failure. He was unable to determine the causes of this, but speculated that alumina and silica impurities somehow affected the grain boundaries. Since his tungsten was ductile, he could simply draw a bar to fine wire, and then winding to filament form was easy. In 1910, the General Electric company began selling such ductile tungsten filaments, and maintained a dominant market position for the next hundred years.

The work of Coolidge is an example of empirical development by careful observation during controlled experimentation - changing parameters one by one and noting effects to develop a working recipe. Ductile tungsten with outstanding resistance to slow straining at high temperature (known as creep) has been steadily improved and understood by studies throughout the entire twentieth century, with further improvements surely to come. It is now understood that doping raw (blue) tungsten oxide with some thousand parts per million of silicon, potassium, and aluminium compounds, then reducing the oxide to tungsten metal, leads to the incorporation of some hundred parts per million of these metals (the rest is lost by evaporation) inside the tungsten powder particles as potassium aluminosilicate. When such metal powders are pressed and sintered (heated to high temperatures for the powder particles to merge), a small amount of elemental potassium is retained, trapped in pores or bubbles inside the fully dense tungsten. The figure at the beginning of this section shows coarse (micron sized) and fine (tenth of a micron) bubbles. When heated to very high temperatures, the potassium evaporates (creating high pressure since trapped inside the pore) pinning grain boundaries which can thus neither move nor slide - a fine grained material is obtained with strong boundaries. These materials are so complicated that it has required years of advanced material characterisation with theoretical studies of nano-bubble evaporation and pinning, as well as process experimentation to reach today's level of understanding and material sophistication. Importantly, it is now known that both bubbles and grains are elongated as fibres during wire drawing and the bubbles can re-spheroidize during high temperature usage, which greatly affects ductility.

Despite such enormous progress in tungsten filament technology, the past years have nevertheless been dominated by the ban and substitution of tungsten incandescent lamps - despite technological improvements these are simply too energy hungry. Nevertheless, the usage of bubble-reinforced tungsten continues to increase with their use as electrodes in other lighting sources and many other power system components. As another example, future fusion reactors - if ever these leave the stage of laboratory curiosity -

will also make major use of these alloys, such as for diverters inside reactor walls to separate fusion products from the hot plasma.

12: Impossible Crystals:

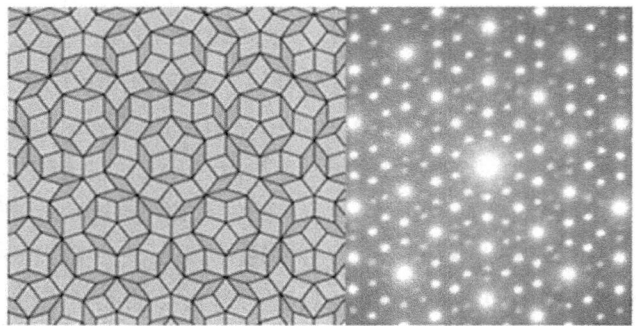

Space Filling and Symmetry

Just over 30 years ago, a scientist at the American National Bureau of Standards noted something so outrageous in an alloy he was studying that it rocked the foundations of the theory of crystal formation based on a two-century-old respected theory. Unable to publish his discovery in the best scientific journals, he was even asked by his laboratory boss to kindly leave. His observations broke unchallengeable rules of crystal symmetry. Yet before even 30 years had passed he had received the Nobel prize in Chemistry for his outstanding determination in convincing the world that his shattering discovery was indeed valid.

The work of Abbé Haüy in the late 18th Century had clearly shown that large, bulk crystals were generated by the repetitive positioning of identical parallelepipeds – made up of individual atoms or molecules and forming unit cell blocks – and the external shape of crystals could be explained by such accumulation. A crystal was defined by the regular order and repetition in three-dimensional space. One important consequence was the presence of translational and rotational symmetry: for example, a cubic crystal looked the same when rotated by 90° about any one of three cube axes, when rotated by 120° about any of four diagonal axes, and so on. All possible crystal structures had been defined within a total of 230 space groups, and it was known that rotational symmetry with 2-fold, 3-fold, 4-fold and 6-fold axes was allowed. Rotation axes of 5-fold, 7-fold, and so on were not permitted. With regards to such symmetrical crystalline structures, mathematical studies had both confirmed the special requirements for space filling and had also suggested that quasi-periodic (or quasi-crystal) structures might exist – showing local ordering and periodicity but not respecting translational symmetry over long distances. As an illustration of this, we can consider the tiling of a bathroom floor with a single tile shape of triangles, squares, rectangles or hexagons – all possible – but try filling the whole floor with pentagons or heptagons and you see the problem! However, the English mathematician Roger Penrose had shown that space could indeed be filled by just two tile forms, rotated and mixed correctly. Such tiling is indeed found covering large wall areas in ancient Islamic palaces. These were still, however, mathematical abstractions and nothing to do with the real world of three-dimensional crystals.

In 1982, on the 8th April to be exact, the Israeli scientist Dan Shechtman, who was spending time at the U.S. National Bureau of Standards, noted something very odd in his laboratory notebook. He was studying a rapidly solidified alloy of aluminium and manganese in the electron microscope (just like a light microscope but using an electron beam instead of light) and noted, with three question marks of surprise, "the sample has 10-fold symmetry???". Expert in the microscope technique, he quickly confirmed his discovery, and also showed that when viewed in different directions the crystal showed sometimes 5-fold, sometimes 3-fold, and sometimes 2-fold symmetry. His sample looked like an icosahedron crystal – a volume contained within 20 triangular faces with 30 edges and 12 corners. He also showed that this apparent shape was not an artefact produced by twinning – a way by which a crystal divides itself into mirror image parts giving falsely-appearing atom distributions. Atoms really had the quasi-crystalline arrangement, considered impossible since it didn't fully fill space.

He shared his results with a colleague Ilan Blech who told him that diffraction patterns like this (arrangement of beams of electrons or x-rays obtained on passing through the crystal and corresponding to the atomic distribution and symmetry) had been seen before but never really explained. Blech used a computer model to simulate diffraction patterns from icosahedra and other non-long-range-ordered quasi-crystals showing that these diffraction patterns could indeed be understood. Shechtman and Blech tried to publish these results in crystal-physics journals, but their work was rejected as incorrect physics, it was perhaps of some value to metallurgists interested in general alloy microstructures and properties. Encouraged by John Cahn, a senior scientist at the National Bureau, Shechtman insisted in trying to publish his studies in a prestigious physics journal, and eventually managed this, leading to much excitement in the scientific community. The work on quasi-crystals demonstrated that these could indeed lead to fully-dense atomic packing even though there was no long-range translational order and had previously-unacceptable symmetries. As a result of this work, in 1992, the International Union of Crystallographers altered the formal definition of what was a crystal to include the options given by the quasi-crystals. Completing this story, in 2011, Shechtman was awarded the Nobel prize in Chemistry for his perseverance in disbelieving the written laws of crystal physics and in demonstrating that his novelty was indeed true.

Since this discovery, quasicrystals are found in many alloys, especially in rapidly solidified aluminium alloys with transition metals like manganese, iron, copper, cobalt, but also in alloys of cadmium, titanium, zirconium, and others. Many quasicrystal types have been discovered, with a variety of symmetries. Initial quasicrystals were discovered using the same rapid solidification method as Shechtman, and all were thought to be metastable, i.e. they would change to a stable crystalline form if heated. Quasicrystals have now been found in a wide range of materials produced using ordinary methods of solidification, offering possibilities of manufacturing large volumes of quasicrystalline materials. They have even been found within inorganic minerals and meteorites, and in nanoscale surface layers, and it has become clear that they are not at all uncommon crystalline materials.

What are the characteristics and typical properties of these quasicrystalline alloys, and what might be applications? The most interesting properties follow from the lack of long-range atomic order on well-defined planes. Long-range order is critical for the easy propagation of heat and electricity (carried by phonons and electrons), as well as plastic deformation (which depends on dislocations moving). Crystals vibrate their atoms as collective excitations down the well-defined crystal planes, and so can transport heat (thermal excitations) and free electrons very well. The lack of long-range crystal planes in the quasicrystal means that phonons move with great difficulty, and thermal conductivity is incredibly poor. It is two orders of magnitude smaller than aluminium and even below that of ceramic zirconia, a commonly used insulator. Electrical conductivity also depends (but only partially) on transport by phonons, and so tends to be low but it varies strongly with temperature. This strange combination of poor thermal conductivity with moderate electrical conductivity makes it an interesting thermo-electric material – creating an electric voltage or current when there is a high temperature gradient. There is strong interest in such materials to generate electricity from spare heat – for example from hot exhaust pipes on cars. Since it is difficult to move dislocations, quasicrystals are hard but also brittle. They have high hardness – difficult to scratch – but cannot stop breaking once a crack forms. High hardness with resistance to chemical attack and low surface energy means that they have low friction and wear coefficients – they are non-stick and non-wear. This has led to their use as coatings on cooking utensils – millions of frying pans are plasma-sprayed each year with CRISTOME coating. Like Teflon, it doesn't stick when food is cooked, but also it doesn't scratch when a knife scrapes out the food. The surface behaviour and good thermal resistance make it an interesting coating for diesel engine protection. Also important is the low surface energy and high thermal stability for quasicrystalline particles in stainless steel to create materials with outstanding mechanical properties and chemical resistance used in medical instruments like ophthalmic needles.

In a short time, the discovery of fundamentally impossible crystalline materials has run through the full course of incredulity, rejection by peers, acceptance, and ovation (unpublishable, to Nobel quality). The theory of crystal structures and symmetries has been rewritten after two centuries. Novel materials are used. An accidental discovery by an insistent scientist.

13: Remembering Shape:

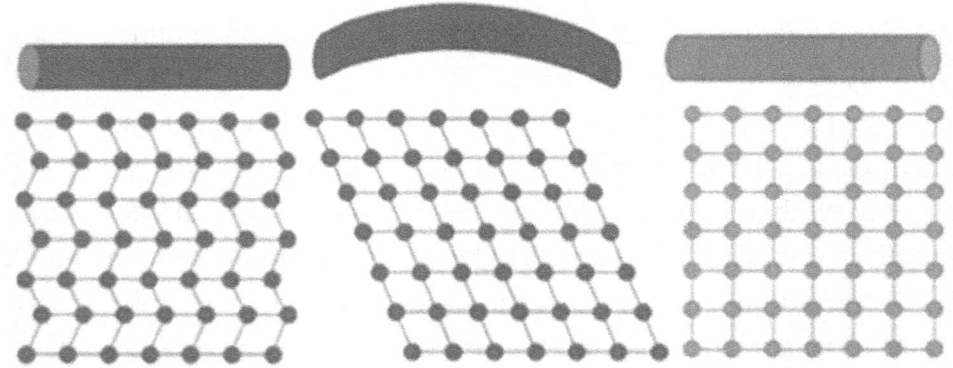

Shape and Phase Change

One of the objectives of this short book is to draw attention to the role of good luck in many cases of materials discovery or development throughout history. These are cases where good luck, or serendipity as it was called by Horace Walpole, Earl of Orford, played a part. However, as Walpole said, such serendipity only works when the mind of the discoverer is already primed and attentive to notice what Mother Nature is going to offer. The present example is one where not only one, but two, lucky events occur and which lead to the discovery of something really strange. In the present case a scientist working in one of the United States military laboratories noticed something strange in the behaviour of some of the alloys that he was investigating, and later one of his bosses did something a little silly, just to see what would happen, and something really strange was found.

The 1950s and 1960s marked a period of great interest in the development of metallic alloys that were relatively light and could withstand high temperatures. These were the years of the space race where rockets and supersonic aircraft were being developed, as well as their high performance jet and rocket engines. In the Naval Ordnance Laboratory in Maryland, the United States, the researcher William Buehler was looking at Nickel-Titanium alloys to be used for making the nose cones of missiles, and these alloys seemed to be ideal for this purpose. These were relatively light materials (titanium has a low density), and were heat-resistant, meaning that both strength and oxidation resistance were maintained to high temperatures. Alloys with approximately equal amounts of nickel and titanium seemed good and showed both high strength and good ductility. These alloys were given the name Nitinol, as an acronym for NIckel-TItanium from the Naval Ordnance Laboratory.

During his experiments in 1959, Buehler made up a series of alloys with a range of nickel and titanium compositions by electric arc melting and casting into bar form. His intention was to examine the role of such changes of composition on material properties. He made up six such alloy bars, one by one, and put each bar on a table to cool as he continued to cast the next bar. At the end, he had his six bars on the table, the first one fairly cool and the others warm or hot. Not waiting for all to cool, he picked up the first, coolest one to grind the surface and remove any surface flaws before moving on to examine the properties. In an accident of fate (or by deliberate intention as Buehler claimed) he dropped the cool bar onto the floor where it made a "flat" noise like a piece of lead – not at all the sharp ring that would be expected of a hard metallic bar!! Was there something wrong with the bar? Quickly Buehler picked up another, warmer bar and threw it on the floor. A good "bell" metallic sound was heard. The next, then the next – all the bars fell like good metals. What was wrong with the first bar? Buehler repeated his tests – picking up and throwing on the floor – and eventually discovered that all bars gave a good metallic "bell" ring when warm, but a dull "thud" when cold. What was happening?

In the meantime, Buehler continued his development of nickel-titanium alloys for the rocket nose cone application, and in 1961 presented his strong, shiny and ductile materials to his supervisors in the laboratory. He had formed the alloys into sheets, and to show how ductile they were he had folded pieces of these sheets into concertina shapes by bending back and for. The pieces were passed around the conference table for the supervisor directors to handle. One of these, a pipe smoker, decided to test the high temperature stability by heating the concertina with his pipe lighter. What was the amazement of all to see the concertina flatten itself to flat sheet when heated! The phenomenon of shape memory had been discovered.

It is now understood that these shape memory materials are capable of transforming their atomic crystalline structure from a symmetric cubic one to a distorted martensitic one by a reaction called a thermoelastic structural change. Changes of elastic properties were reported within a gold-cadmium alloy by Ölander in Stockholm in 1932, and researchers in the United States reported in 1938 the appearance, then disappearance, of martensite in a brass alloy as the temperature fell, then rose. Studies by Russian scientists Kurdjumov and Khandros over the next decade laid the groundwork for understanding these thermoelastic phase changes - a change of crystalline structure by co-ordinated atomic shuffles that depended on applied stress and temperature but required no atomic diffusion - they were instantaneous reactions that did not require time for the transformation to occur.

At low temperatures in these materials, the atomic arrangement is called martensite, where the atoms are arranged in a distorted (non-cubic) manner. Different regions of an object take up different directions of distortion such that the overall shape is maintained. When a load is applied, the material shows an enormous amount of deformation by the growth of some of the distortions while consuming unfavourable variants. Imagine that cubes are distorted (to diamond shape) either to the left, the right, up or down – if I load to the right, the right distorted variant grows while the left distorted variant shrinks. If I take off the load, the object returns to its original shape as the variants return to the original mixture. This is called the superelastic effect, and the amount of elastic (i.e. recoverable) deformation is enormous compared with all normal metals. If I continue my experiment and deform the material much more, the shape change is retained when I unload – if I bend a wire, it stays bent. If I now heat the sample in its distorted shape, the atoms take up the second, cubic arrangement, known as austenite, and as the martensite variants are lost the material returns to its original shape. In the figure above, the original material at left has two martensite variants, deforming the wire leads to a predominance of one variant, while heating to cubic austenite recovers the original straight wire. So, for example, a tube can be flattened when cold and will recover the tube shape when heated. By very carefully training material both when cold and when hot, imposing a special mixture of martensite variants at the low temperature, it is possible to train what is called two-way memory whereby one shape is held (remembered) at low temperature, a different one remembered at high temperature, and the material can be cycled back and forth between the two shapes by changing temperature up and down. The temperatures where the changes occur are fairly low, for example at or below room temperature for changing to martensite and 100-200°C for changing to austenite. This is the approximate temperature range where shape changes can be remembered.

Since the discovery of Nitinol, several other shape memory alloys have been developed for commercial use – especially important are alloys with composition about 55%Ni-45%Ti, Cu-15%Al-4%Ni and Cu-25%Zn-4%Al. These generally use the shape memory effect, but sometimes it is the superelastic effect that is important. In general engineering, hydraulic pipe connectors and water-sprinkler alarm systems exist using the shape memory effect – a connecting ring of expanded diameter or a connecting spring changes size or shape when temperature changes. Surgeons insert blood-clot filters or artery stabilising stents in collapsed form into the body for the filter or pipe to expand to proper shape once heated by the body. Eyeglass frames can be distorted superelastically by sitting on, but return afterwards to shape, and orthodontic wires stretch superelastically to retain high correcting loads on teeth as these grow back to their correct positions. A recent application creating some excitement is the use of shape-changing chevrons at the rear of aircraft engines – these direct the exhaust flow when warm (at ground temperature) to reduce noise for take-off and landing, but retract away for higher engine efficiency in the cold of high altitude.

From accidentally dropping samples on the floor, to checking out behaviour with a pipe lighter, a family of new alloys has developed that play an important role in many niche high-technology applications.

14: Non-Crystalline Metals:

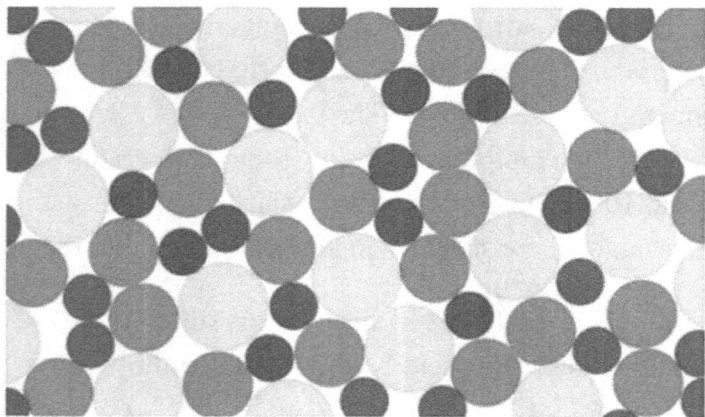

Non-crystalline atomic arrangement

The word glass is usually associated with windows. Since the beginning of civilization a glass was a ceramic material, usually a mixture of oxides of metals like silicon, sodium or calcium, that could be heated to flow like a thick liquid, was more or less transparent, and was an inorganic insulator. The materials were all non-crystalline, that is that the atoms were arranged

more or less at random (as in the picture above) and not in the regular crystalline arrangement found for all metals and their alloys. In fact it was "known" that metals like the crystalline state so much that they could not be made glassy. Then, in 1960, a Belgian scientist, Paul (Pol) Duwez, working in the California Institute of Technology reported that he had made a gold-silicon alloy amorphous (or non-crystalline), and shortly afterwards showed how simple steel-like alloys could be made with this "impossible" glassy structure.

Duwez had been working in the American Jet Propulsion Laboratory on possible new high-temperature alloys and ceramics to be used for rocket nozzles. These were the years of the second world war and the following cold war, so rocket science was new and important. In 1959 he began using the so-called splat-quenching method to cool materials extremely rapidly from the liquid state. After first studying alloys of simple metals like copper and silver, and obtaining increases in the amount of intermixing (solubility) in his solid samples – like mixing more sugar into hot coffee than into cold coffee – he moved on to mixtures containing more complex metals like silicon into gold. Amazingly for the time, he found that he had made a glassy material. He continued his study to look for magnetic glasses and, using the hypothesis that it was better to mix metal and non-metal elements, he began to study Fe-P alloys. His student at that time, Lin, had forgotten that he should not melt this alloy in a graphite crucible since the alloy would react and take up a lot of carbon, and so he accidentally processed a Fe-C-P alloy – and easily obtained metallic glasses. They were even magnetic, as desired. In fact, as further studies showed, it was not possible to obtain metallic glasses with binary Fe-P alloys, and it was the lucky mistake of the research student that opened up this important field of study.

The splat-quenching technique involves taking a small droplet of hot liquid metal and smashing it quickly between two cold copper plates – the droplet is smashed to a very thin disk, some tens of microns thick, and cooled at a speed of about one million degrees per second. This is so rapid that the atoms do not have enough time to rearrange themselves from the disordered atomic state of the liquid into a complex crystalline arrangement, and so glassy materials can be made. In fact all the simple metals (like aluminium, copper, gold, iron and so on) still have plenty of time to form crystals, but alloys with complicated chemical compositions that are mixtures of metals and non-metals (like boron, carbon, silicon) do not have enough time to separate chemically and form the complex crystalline structures of any possible solid phase. Hence these alloys become glassy. This idea, of mixing dissimilar atoms that would form complex crystals in equilibrium, became the basis of scientific theories to help find new glassy alloys. At the same time, it is obvious that the faster cooling that is managed, the less time that will be available for atomic re-arrangement into crystalline form and the greater the chance of making a glassy material.

Since these revolutionary experients at the beginning of the 1960's, a variety of techniques has been developed to produce commercially-valuable quantities of rapidly quenched glasses. All techniques involve extracting heat quickly from a thin layer or jet of molten alloy, and so can produce powders, or wires, or long, thin sheets of material. Tons of such material are produced commercially each year, such as the so-called Metglas alloys. In parallel, the theories of alloy crystallization have given insight to which useful compositions might be quenched with very slow cooling rates and have led to the development, in the 1990's, of the so-called bulk metallic glasses, such as the Zr-Ti-base Vitreloy with sizes of several centimetres.

While traditional metals and alloys all have the crystalline state, that is that the mixtures of atoms adopt the regular characteristic atomic arrangement, the metallic glasses essentially have the disordered atomic arrangement found in the liquid state. The preparation technique quenches or freezes this arrangement to low temperatures where further atomic movement becomes too slow to notice. The typical properties are a consequence of the uniform chemistry and absence of crystalline defects – quenching did not allow the separation of atoms over distances of even a nanometre in the glass, while the conventional crystalline material will typically contain micron-sized grains of different chemical composition, a variety of grain and phase boundaries, as well as many planar and line defects. Being metal-based, they possess electrical and thermal conductivity, but not generally as good as found in simpler metallic alloys. At the same time, since the resistivity is so high, it changes little with temperature. When the majority element is magnetic (for example iron or cobalt) the glasses will also be highly magnetic (easily magnetised to high levels of magnetism) and have low coercivity (it is easy to magnetise or reverse-magnetise when the applied magnetic field reverses). The glasses have interesting mechanical properties, controlled by the absence of the defects found in conventional materials and which define their strength and toughness. Such defects allow easy plastic deformation (make the crystalline material soft), while the metallic glass is typically extremely strong – for example Fe-base glasses can be as strong as the strongest piano-wire steels. Unfortunately these same defects in the crystalline materials are useful in delaying fracture, and thus allow us to obtain tough materials. Without these defects, the metallic glasses are left with very little resistance to fracture and break easily.

In view of their strange properties, metallic glasses have found a number of specialist (niche) applications. The low coercivity and high resistance means that there are only very low energy losses as the magnetic field switches from up-to-down. and back, inside special high-efficiency electrical-voltage transformers. High, but constant electrical resistance means that they can be used in electronic circuits where constant behaviour is important, even though the temperature changes. Titanium-base Vitreloy bulk glasses are very strong and wear resistant, with low elastic modulus and good biocompatibility, such that they may be used as implants for bone repair or replacement. In a similar way, some magnesium-base bulk glasses with good mechanical properties may be used for temporary bone repair, since they will slowly dissolve in the body and disappear once the body has repaired itself. A variety of other applications are still under study.

This is yet another example of how established scientific theory flies out the window – metals and alloys were "known" not to exist with the disordered, non-crystalline atomic structure. Again, scientific experimentation into other phenomena – increasing the amount of solubility in the crystalline state – led to the discovery of "impossible materials". Yet again, it is only when the young research scientist makes an experimental mistake – using the unsuitable crucible equipment – that another breakthrough observation is made. So much of our scientific cleverness is really due to error and luck.

15: Shock-proofed ceramics:

Shock-proof oven dishes

Ceramics are known to be hard materials, but brittle, and break if hit with a hammer or dropped on the floor. Our everyday experience tells us this – that window glass breaks when hit, a pot or dish shatters on falling to the floor. Even the thermal shock of putting a cold dish into a hot oven or microwave heating is too much for many such ceramics. At least it was until a superbly tough shock-proof ceramic was discovered, by an accident of serendipity, in a glass laboratory interested in other properties.

Dr S. Donald Stookey had studied chemistry and researched his thesis in MIT, before deciding to join the Corning Glass Works in 1940 in New York State to look at the structure and properties of glasses. This seemed far away from chemistry but closer than the study of baking, which was the other job offer made to him. Dr Stookey saw himself as more like an alchemist that a scientist, since glasses were so little understood at that time. It was known that glasses were like frozen liquids, with unstable non-crystalline structures, and that doping and annealing could lead to the appearance of minute crystals which would change the glass properties.

The 17th century had seen the development of a formula and process for preparing gold-ruby glass. It had been believed that cups of this material could transmit the powers of the philosopher's stone – the drinker would imbibe the elixir of life. Stookey realised that the wonder of glass lay in the flexibility offered by doping with suitable seed materials, followed by heat treatments to grow these seeds to the required crystals. Doping with traces of copper or gold led to the famous ruby glass. Doping instead with fine particles of a fluoride led to light scattering and an opal (milky) appearance. Stookey developed what would be called the FotoForm glass – a glass sensitive to ultraviolet light where nuclei could be excited by shining ultraviolet light and then grown by subsequent heating. This material had an enormous variety of uses, all involving freezing an optical image in a glass as an array of microscopic crystals, or perhaps etching the crystals away with acids. A first demonstration by Stookey made a paperweight with a photograph of his wife captured inside. He suggested that holograms could be frozen in the same way; the US Treasury could make unforgeable money as glass pieces containing delicate images; spies could pass on hidden messages inside glass which would be heat treated by the receiver; glass with thousands of close holes could be used to guide electron beams. Such developments were keeping Stookey busy, and his company happy, with dozens of new materials, applications and businesses.

Then, one day in 1953, serendipity struck. Stookey put his experimental glass into his furnace for the usual 600°C soak. This would grow the usual fine and delicate crystals and develop the optical properties. Unfortunately (!) that day his furnace thermometer stuck at an excessively hot 900°C, and he over-cooked his material. The story goes that he angrily pulled the object out of the furnace with his tongs and either it fell, or in a rage he threw it, onto the floor where it bounced with a loud clang. The glass had gone a deep milky texture and was so tough that it could only be broken with the most intense of efforts. Pyroceram had been invented. A few years later this appeared on the mass market as CorningWare used for ovenware, as shown in the picture at the beginning of this chapter. At one time it was claimed that every household in America possessed dishes made of this material.

The crystals formed at the much higher temperatures were different and grew to become much larger than the fine ones desired for the photographic applications. Most importantly, the growth of these crystals had generated incompatibility internal stresses inside the glass and these meant that any crack trying to grow from the outside surface would be blunted and stopped before growing across the entire object. The effect is very much the same as steel reinforcing bars placed inside concrete which prevent the brittle concrete from falling apart. These internal stresses made the material virtually unbreakable.

Over the subsequent years a large number of pyroceramic varieties were developed. Not only did these materials resist cracking and were very hard and scratch-proof, they were resistant to high temperatures (up to 900°C), and some had low expansion coefficients and were transparent to radar electromagnetic waves. They were suitable for use as radar domes and nose cones of military missiles, and were the basis for the development of Gorilla Glass, now used in iPhones and in many Liquid-Crystal-Displays.

Final Comments

I hope that you will have enjoyed reading these short descriptions of how materials have been discovered and developed over the last millenia. I hope also that you will have learned more about the science of materials.

In some of the cases that I have discussed, it has simply been Nature that offered to us the best materials for us to discover and put to good use (Flintstones, Fluid Bronzes). In other cases, mankind has improved the materials that are available by using good common sense. In many of these cases, however, he has been helped by lucky accidents or mistakes that have brought to his attention things that would otherwise have remained hidden from view (Age Hardening Aluminium, Remembering Shape). On some occasions, these discoveries have rocked what we had thought was good understanding of the basic physics behind the material (Impossible Crystals, Non-Crystalline Metals).

All useful materials are – as the adjective says – useful for human society in some way. Many materials have been developed with a specific application in mind, and the accidental discoveries have improved the applications or made new ones possible. In some of the cases described, as the title of this work has implied, the new material or the properties of the material used have had such an impact that they have changed the course of our history (Napoleon's Buttons, Nelson's Bottoms). In all of the cases described here, the new materials have had a major effect on our lives – by providing us with efficient jet engines (Superalloys), with clean knives for the kitchen (Stainless Steel), with bright house lighting (Let the Light Shine – Tungsten Filaments), and so on.

In a world where everything seems so carefully controlled, especially the behaviour of materials and engineering devices, it may come as a surprise to realize that many major discoveries have relied so much on good luck. This has been the case over the millenia, but remains true even in today's society where everything seems to be so logical and well planned.

Acknowledgements:

I am grateful to the following sources for illustrations found at the beginning of each chapter. Other illustrations are taken from material in the public domain and from own studies by the author.

Chapter 3: Flintstones:

"Spark from lighter flint". Via Wikipedia
http://en.wikipedia.org/wiki/File:Spark_from_lighter_flint.jpg#mediaviewer/File:Spark_from_lighter_flint.jpg

Chapter 5: Age Hardening Aluminium:

"Interstitial solute" by Interstitial_solute.png: Siamrut at English Wikipedia / *derivative work: Zerodamage - This file was derived from: Interstitial_solute.png. Licensed under Creative Commons Attribution-Share Alike 3.0 via Wikimedia Commons - http://commons.wikimedia.org/wiki/File:Interstitial_solute.svg#mediaviewer/File:Interstitial_solute.svg

"Substitutional solute" by Substitutional_solute.png: The original uploader was Siamrut at English Wikipediaderivative work: Zerodamage - This file was derived from:Substitutional_solute.png. Licensed under Creative Commons Attribution-Share Alike 3.0 via Wikimedia Commons - http://commons.wikimedia.org/wiki/File:Substitutional_solute.svg#mediaviewer/File:Substitutional_solute.svg

Chapter 6: Fluid Bronzes:

© 2004-2013 University of Cambridge; (CC BY-NC-SA 2.0 UK) Creative Commons allows reproduction with appropriate citing. Taken from a Teaching and Learning Package on Solute Partitioning, http://www.doitpoms.ac.uk/tlplib/solidification_alloys/solute_partitioning.php

Bronze Swords:

"Apa Schwerter". Licensed under Creative Commons Attribution-Share Alike 3.0 via Wikimedia Commons - http://commons.wikimedia.org/wiki/File:Apa_Schwerter.jpg#mediaviewer/File:Apa_Schwerter.jpg

Bronze Axes

"Palstave one ring" by José-Manuel Benito Álvarez —> Locutus Borg - Own work. Licensed under Public domain via Wikimedia Commons - http://commons.wikimedia.org/wiki/File:Palstave_one_ring.jpg#mediaviewer/File:Palstave_one_ring.jpg

Chapter 7: The Discovery of Graphene:

Taken from "Graphen" by AlexanderAlUS - Own work. Licensed under Creative Commons Attribution-Share Alike 3.0 via Wikimedia Commons - http://commons.wikimedia.org/wiki/File:Graphen.jpg#mediaviewer/File:Graphen.jpg

Chapter 8: Invariable Dimensions:

Taken from "Invar-Graph-CTE-composition" by RicHard-59 - Own work, based on png-version. Licensed under Creative Commons Attribution-Share Alike 3.0 via Wikimedia Commons - http://commons.wikimedia.org/wiki/File:Invar-Graph-CTE-composition.svg#mediaviewer/File:Invar-Graph-CTE-composition.svg

Chapter 9: Combinatorial Materials:

"6sided dice" by Diacritica - Own work. Licensed under Creative Commons Attribution-Share Alike 3.0 via Wikimedia Commons -

http://commons.wikimedia.org/wiki/File:6sided_dice.jpg#mediaviewer/File:6sided_dice.jpg

Chapter 10: Stainless Steel:

From "Harry Brearley" by David Morris - From geograph.org.uk. Licensed under Creative Commons Attribution-Share Alike 2.0 via Wikimedia Commons - http://commons.wikimedia.org/wiki/File:Harry_Brearley.jpg#mediaviewer/File:Harry_Brearley.jpg

"Kitchen knives" by Themightyquill - Own work. Licensed under Creative Commons Attribution-Share Alike 3.0 via Wikimedia Commons - http://commons.wikimedia.org/wiki/File:Kitchen_knives.svg#mediaviewer/File:Kitchen_knives.svg

Chapter 11: Let the Light Shine – Tungsten Filaments:

"Thermionic filament" by Created by Deglr6328, uploaded by Superclemente - en.wikipedia.org. Licensed under Creative Commons Attribution-Share Alike 3.0 via Wikimedia Commons - http://commons.wikimedia.org/wiki/File:Thermionic_filament.jpg#mediaviewer/File:Thermionic_filament.jpg

Chapter 12: Impossible Crystals:

"Penrose Tiling (Rhombi)" by Inductiveload - Own work. Licensed under Public domain via Wikimedia Commons - http://commons.wikimedia.org/wiki/File:Penrose_Tiling_(Rhombi).svg#mediaviewer/File:Penrose_Tiling_(Rhombi).svg

Taken from "Zn-Mg-HoDiffraction" by Materialscientist - Own work. Licensed under Creative Commons Attribution-Share Alike 3.0 via Wikimedia Commons - http://commons.wikimedia.org/wiki/File:Zn-Mg-HoDiffraction.JPG#mediaviewer/File:Zn-Mg-HoDiffraction.JPG

Chapter 13: Remembering Shape:

Taken from "Constitutive Modeling of Shape Memory Alloys", Research Topic, by the Computational Mechanics and Advanced Materials Group, Department of Civil Engineering and Archtecture, of the University of Pavia, Italy.

Chapter 15: Shock-proofed ceramics:

Taken from Wikimedia Commons. File: Corningware casserole dishes; Reference Author – User:Sparkla; Source http://en.wikipedia.org/wiki/File:Corningware_(flower-print_casserole_dishes).jpg

Notice:

While all attempts have been made here to be as scientifically rigorous as possible with all statements, some slight poetic license may have been taken occasionally and some of the statements may not be exactly correct in all the fine details.

www.ingramcontent.com/pod-product-compliance
Lightning Source LLC
Chambersburg PA
CBHW070920180526
45168CB00005B/2089